curious about
YouTube
AF575624
BY RACHEL GRACK
AMICUS

What are you

Search
YouTube Kids
Google LLC
GET
4+
No16
4.8
An app made just for children.
Recommended

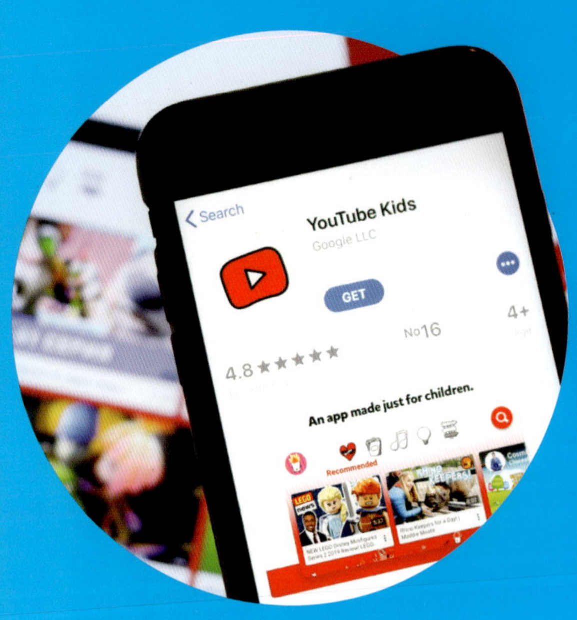

curious about?

CHAPTER THREE

Bet You Didn't Know...

PAGE
14

Curious About is published by Amicus
P.O. Box 227
Mankato, MN 56002
www.amicuspublishing.us

Editor: Alissa Thielges
Series and Book Designer: Kathleen Petelinsek
Cover Designer: Lori Bye
Photo Researcher: Omay Ayres

Library of Congress Cataloging-in-Publication Data
Names: Koestler-Grack, Rachel A., 1973– author.
Title: Curious about YouTube / Rachel Grack.
Description: Mankato, MN: Amicus, [2024] | Series: Curious about favorite brands | Includes bibliographical references and index. | Audience: Ages 6–9 | Audience: Grades 2–3 | Summary: "Nine kid-friendly questions give elementary readers an inside look at YouTube to spark their curiosity about the brand's history and users. A Stay Curious! feature models research skills while simple infographics support visual literacy"—Provided by publisher.
Identifiers: LCCN 2022032135 (print) | LCCN 2022032136 (ebook) | ISBN 9781645493310 (library binding) | ISBN 9781681528557 (paperback) | ISBN 9781645494195 (ebook)
Subjects: LCSH: YouTube (Electronic resource)—Juvenile literature. | Internet videos—Juvenile literature.
Classification: LCC TK5105.8868.Y68 K64 2024 (print) | LCC TK5105.8868.Y68 (ebook) | DDC 384.3/8—dc23/eng/20221018
LC record available at https://lccn.loc.gov/2022032135
LC ebook record available at https://lccn.loc.gov/2022032136

Photos © Alamy/MediaPunch Inc 16–17, NetPhotos 14–15, WENN Rights Ltd 11, ZUMA Press, Inc. 4–5; Dreamstime Jenya Pavlovski 21 (tl circle), Mykhailo Polenok 13 (t), Transversospinales 7; Flickr/Wikimedia Commons/Like Nastya 13 (b); Kathleen Petelinsek 22 and 23; Shutterstock Bricolage 21 (t), Chaay_Tee 21 (bl circle), DisobeyArt 21 (tr circle), Faizal Ramli 9 (2), Media Home 21 (br circle), Monkey Business Images 6, PO11 9 (1), ZikG 18–19; Wikimedia Commons/Asylum Records/Atlantic Records 9 (4), Atlantic Records Press Release 9 (5), Jawed Karim and Yakov Lapitsky 8

Printed in China

CHAPTER ONE

Who started YouTube?

DID YOU KNOW?
Over 30 million people visit YouTube every day.

Left to right: Steve Chen, Chad Hurley, and Jawed Karim

Three friends: Chad Hurley, Steve Chen, and Jawed Karim. They wanted to share their videos with each other. So, they created YouTube. In 2005, they **uploaded** the very first video. The site became wildly popular. It is the top video-sharing **brand** in the world!

What can you watch on YouTube?

Kids can watch safe videos just for them on YouTube Kids.

Pretty much everything! Many people like to watch music videos and funny clips. Today, Google owns YouTube. It's easy to find videos on how to do almost anything. You can even rent TV shows and movies. YouTube also has live TV for sports and TV shows. But you must pay a monthly **fee**.

DID YOU KNOW?

People used to call a TV "the tube." YouTube stands for "your TV." It's a show where you are the star!

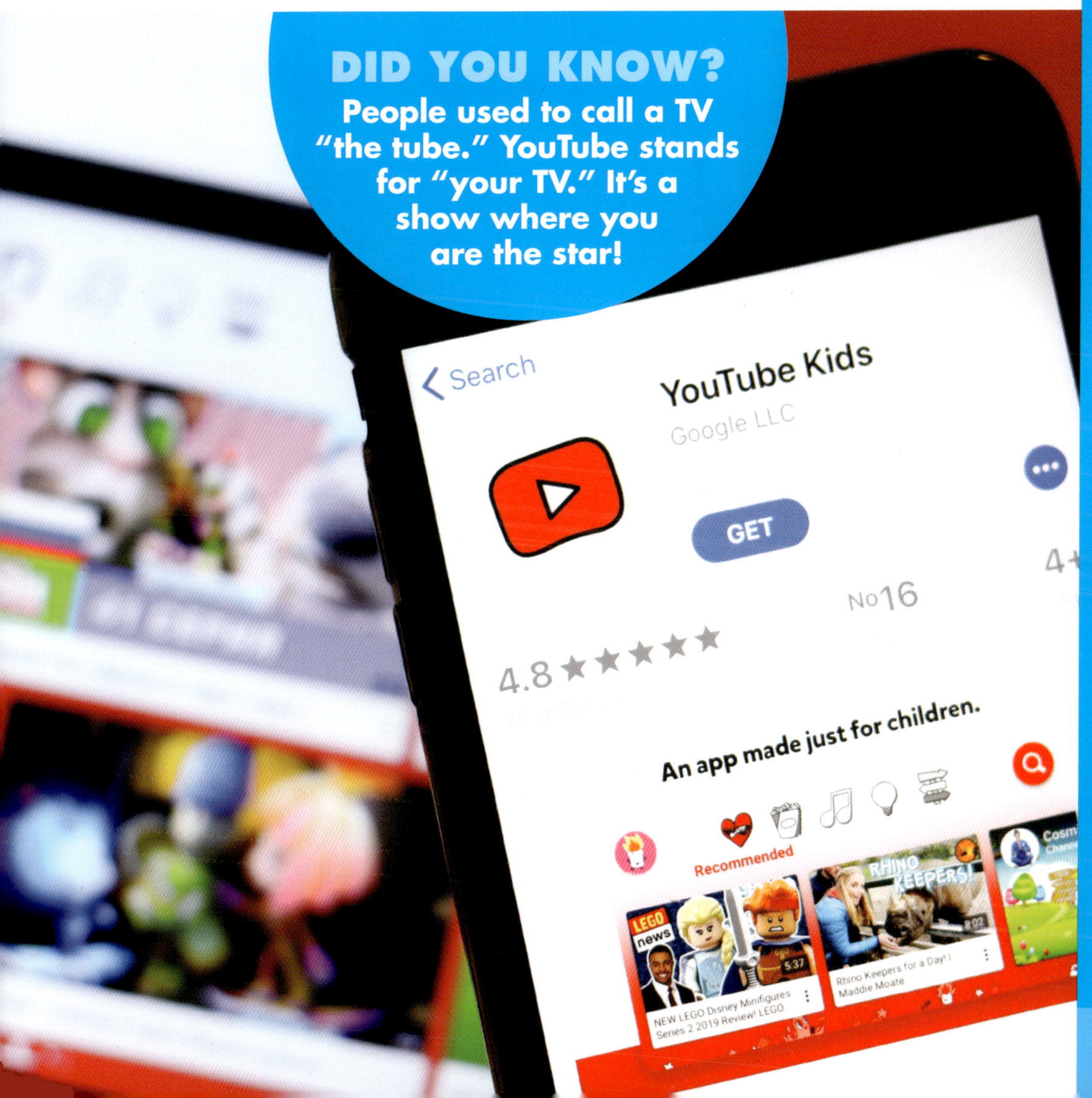

An app made just for children.

CHAPTER TWO

What was the first YouTube video?

"Me at the Zoo" is still on YouTube.

"Me at the Zoo." In it, Jawed talks about elephants. It is only 19 seconds long. At first, YouTube had few **hits**. That changed in December of 2005. Someone posted a video clip from a popular TV show. It got over five million views within a few days. YouTube went **viral**.

1

BABY SHARK DANCE
11.7 BILLION VIEWS

2

DESPACITO
8 BILLION VIEWS

3

JOHNNY, JOHNNY, YES PAPA
6.5 BILLION VIEWS

4

SHAPE OF YOU
5.8 BILLION VIEWS

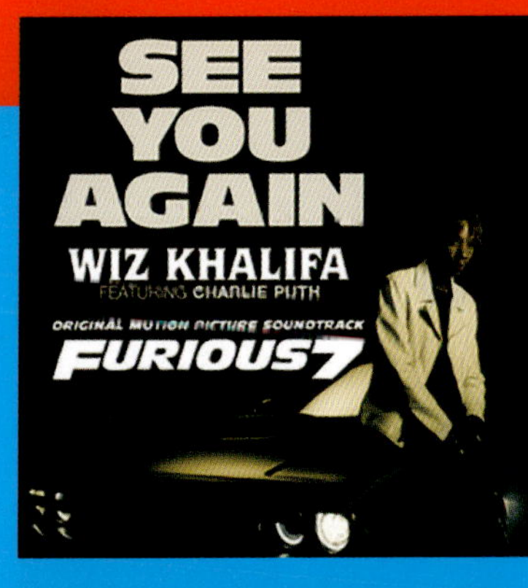

5

SEE YOU AGAIN
5.7 BILLION VIEWS

How does a video go viral?

It takes two things. One, the video draws more than five million views. And two, it happens in a week. Most viral videos are funny, interesting, or thoughtful. Anyone can make one. No special skills are needed. Often, it's just great timing. Be ready to record at any time!

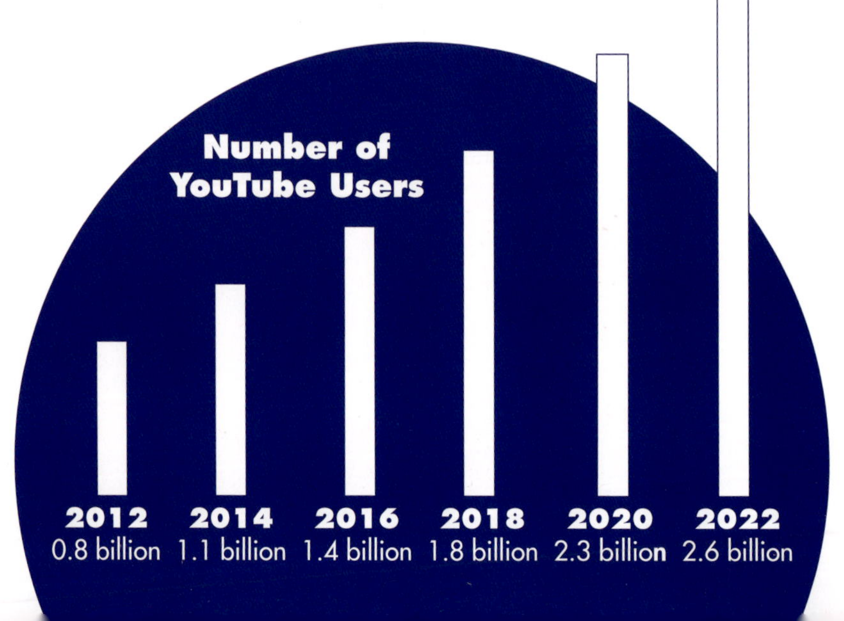

These girls sang a Nicki Minaj song that went viral. It has 55 million views!

What are YouTubers?

You probably know a few. Maybe it's you! YouTubers upload videos regularly to their own **channels**. Their videos often have a **theme**. Some make funny comments while **gaming**. Others review new products. Kids are famous YouTubers, too. Their parents often help them. People **subscribe** to watch their favorite YouTubers.

DID YOU KNOW?

The longest YouTube video is 23 days long! The shortest is less than one second.

Nastya and her dad make videos every week.

CHAPTER THREE

How does YouTube know what videos I like?

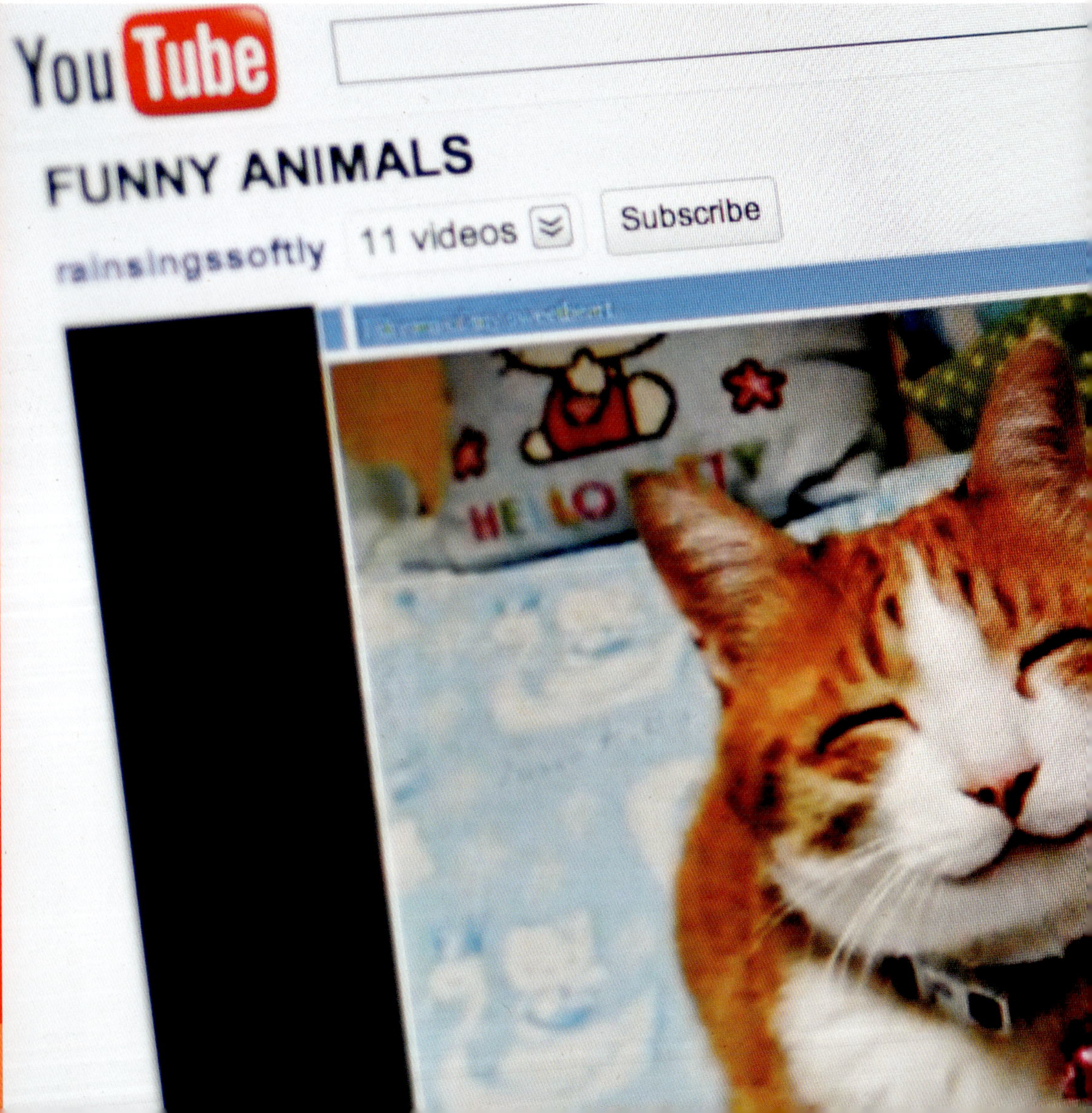

It tracks what you watch. Do you watch unboxing videos? It will show you more. YouTube also **recommends** videos other people have liked. You may like the same stuff. Popular videos are shown more often.

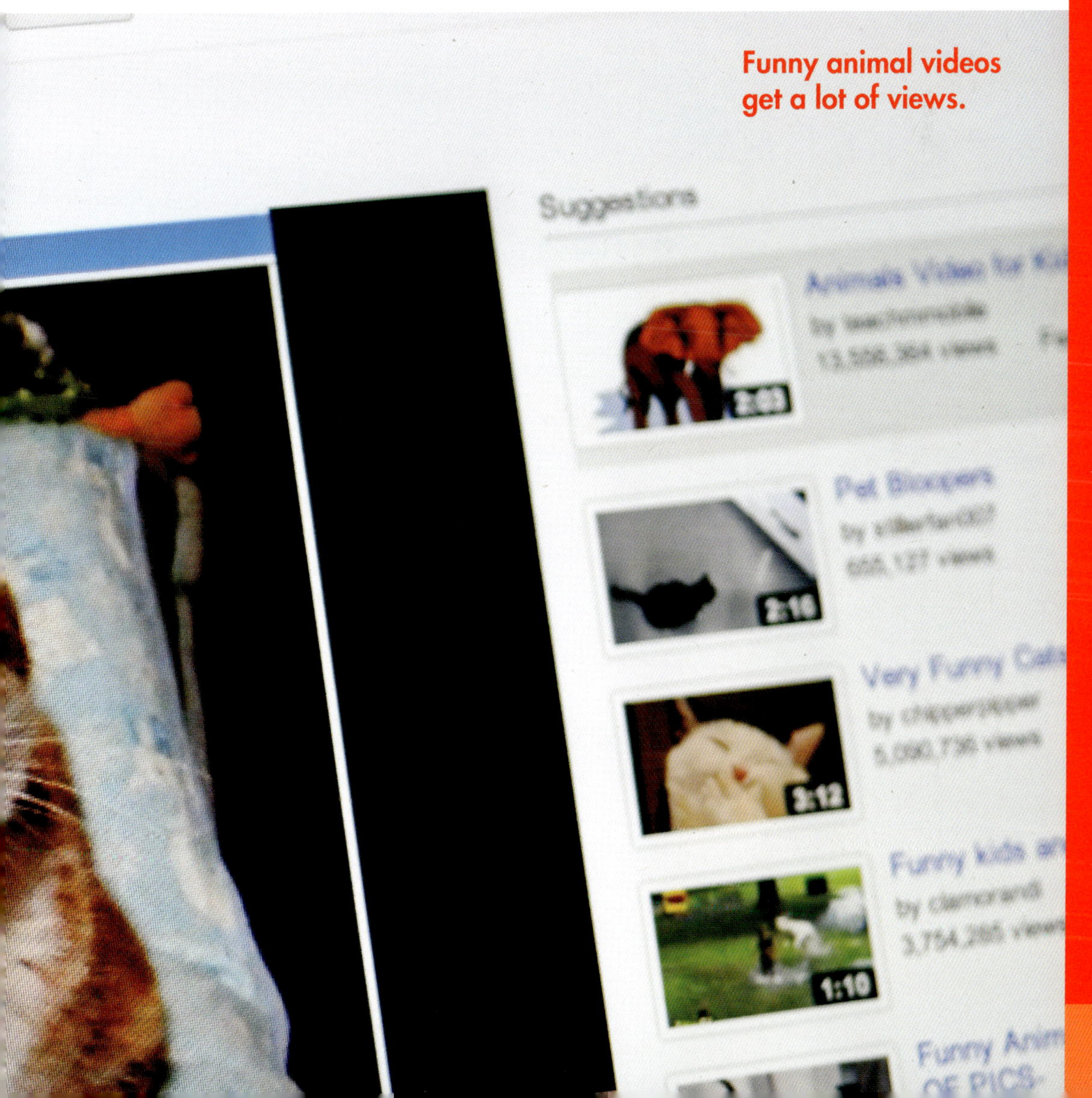

Funny animal videos get a lot of views.

What is the most popular channel?

Dude Perfect even does live shows.

T-Series. It's a music company from India. It had the most subscribers in 2022. But Dude Perfect is one of the most watched. This group does trick shots and funny challenges. Many kids know Ryan's World and Like Nastya. They are the most popular kids' channels.

Do all YouTubers get rich?

DID YOU KNOW?
Each click of an ad is worth 18 cents. YouTubers need a lot of ad clicks to make millions!

Most do not. But some make millions of dollars each year. They get paid through ads. A channel needs at least 1,000 subscribers first. Some YouTubers charge a fee to watch. Many also create and sell their own **merchandise**.

Ryan earns some money from toys he rates on his channel.

Could I be a YouTuber?

Of course! All you need is a video recorder and a YouTube channel. It also helps to have a cool idea and a fun personality. Feel nervous in front of a camera? Team up with a friend. Try different ideas. Stick with what gets more views. You could become the next YouTube star!

POPULAR YOUTUBE IDEAS

- Silly (but safe!) stunts
- Anything with food
- How-to videos
- Toy reviews
- Challenges
- Gaming

STAY CURIOUS!

ASK MORE QUESTIONS

What kinds of YouTube videos do kids make?

Which YouTubers have gaming channels?

Try a BIG QUESTION: How could I get more YouTube views?

SEARCH FOR ANSWERS

Search the library catalog or the Internet.
A librarian, teacher, or parent can help you.

Using Keywords
Find the looking glass.

Keywords are the most important words in your question.

If you want to know:

- about kids on YouTube, type: CHILD YOUTUBERS
- for gaming channels, type: YOUTUBE GAMERS

LEARN MORE

FIND GOOD SOURCES

Are the sources reliable?
Some sources are better than others. An adult can help you. Here are some good, safe sources.

Books

YouTube
by Emma Huddleston, 2019.

YouTube
by Lisa Owings, 2017.

Internet Sites

Kiddle: YouTube Facts for Kids
https://kids.kiddle.co/YouTube
Kiddle is an online encyclopedia for kids.

YouTube Kids
https://www.youtubekids.com/
YouTube Kids is a kid-friendly YouTube app safer for kids to search video content on their own.

Every effort has been made to ensure that these websites are appropriate for children. However, because of the nature of the Internet, it is impossible to guarantee that these sites will remain active indefinitely or that their contents will not be altered.

SHARE AND TAKE ACTION

Think of ideas for a YouTube channel.
Work with a friend. Test your ideas by recording short videos. Then show your videos to friends and family. Use their feedback to improve your YouTuber skills.

Is there something you would like to learn how to do?
Ask an adult to help you search YouTube for a how-to video. Did the video help you?

Ask your teacher if your class can make a YouTube video of a science experiment.

GLOSSARY

brand A group of products made or owned by the same company.

channel A YouTube page with uploaded videos.

gaming Playing video games.

fee A charge for a service.

hits Visits to a website.

merchandise Goods that are bought and sold.

recommend To suggest as being good.

subscribe To sign up to receive the latest videos of a YouTube channel.

theme The main idea.

upload To transfer a video from a recorder onto YouTube.

viral Quickly spread across the internet.

INDEX

About the Author

Rachel Grack has been editing and writing children's books since 1999. She lives on a small ranch in southern Arizona. As a lover of stories, Disney is close to her heart. She also enjoys watching her grandchildren play Pokémon games on her Nintendo 64 console.